# Separations Division 2023

AA008717

Held at the 2023 AIChE Spring Meeting and 19th Global Congress on Process Safety

Houston, Texas, USA
12-16 March 2023

ISBN: 978-1-7138-7925-1

**Printed from e-media with permission by:**

Curran Associates, Inc.
57 Morehouse Lane
Red Hook, NY 12571

**Some format issues inherent in the e-media version may also appear in this print version.**

Copyright© (2022) by AIChE
All rights reserved.

Printed with permission by Curran Associates, Inc. (2023)

For permission requests, please contact AIChE
at the address below.

AIChE
120 Wall Street, FL 23
New York, NY 10005-4020

Phone: (800) 242-4363
Fax: (203) 775-5177

www.aiche.org

**Additional copies of this publication are available from:**

Curran Associates, Inc.
57 Morehouse Lane
Red Hook, NY 12571 USA
Phone: 845-758-0400
Fax: 845-758-2633
Email: curran@proceedings.com
Web: www.proceedings.com

# TABLE OF CONTENTS

## 24b Column Inspections and Safety

*Tony Cai*

Distillation column inspections are an essential step to a successful installation or revamp. Safety, mechanical design and knowledge of the process intended for the column are just a few of the topics that must be covered before entering the column. Setting up a checklist of items to be checked alone often requires significant attention to detail. This presentation will provide an introduction to preparation for a safe and successful column inspection.

## 24c Ideal Stage Vs. Rate-Based Simulation of Gas Treating Processes

*G. Simon A. Weiland*

Rate-based simulation for sour gas treating applications is a relatively new, but proven, technology that was developed in the 1980s and 1990s. The older more traditional technology uses the concept of the ideal and has been used to simulate acid gas removal applications throughout the industry for many years, so what are the advantages to using rate-based simulators? The discussion will cover the conceptual differences and technical bases between mass transfer rate-based simulators and ideal-stage based simulators, why someone would choose one over the other, and the advantages and disadvantages of each.

## 24d Artificial Intelligence Applications in Chemical Process Industry

*Zheyu Jiang*

In the era of Industry 4.0, thanks to the availability of big data and computational capabilities, the chemical process industry is embracing broad adoption of artificial intelligence (AI) and machine learning (ML) methods. In this talk, we offer a holistic view of how AI transforms chemical process industry at scale. We start by presenting a historical perspective on how chemical process industry developed and adopted computer-aided decision-making tools to synthesize, design, optimize, and control process systems. Next, we highlight some of the state-of-the-art R&D efforts and applications in AI that address critical problems in chemical process industry, including predictive analytics (e.g., material properties prediction and new material discovery), process monitoring and quality/safety assurance (e.g., fault detection and diagnosis), as well as process optimization and control (e.g., supply chain optimization, optimization of batch and continuous manufacturing processes). Finally, we envision the ever-growing role AI can play in chemical process industry in the future and discuss how the industry should reshape its culture to prepare for this trend.

## 24e Troubleshooting Distillation Columns

*Andy Burleigh*

In all chemical plants and refineries there generally is not enough instrumentation in the right places in our processes. During the design phase, decisions were made between knowing more about the process, which means buying more instruments, and saving capital funds. When a problem appears with the operation of a vessel, there are always gaps in the knowledge of what is happening due to a lack of instruments in the places where knowledge is not routinely required. Process engineers and operations personnel need to know what is happening so they can take corrective action. When knowledge is missing, they often take actions that make matters worse. For distillation and separation processes Chemical Engineers have historically relied upon plant process measurements such as low rates, temperatures, and pressures and model or simulation results for data with which to solve troubleshooting or re-design problems. An array of on-line diagnostic services are available that can provide additional data offering real-time information on how pieces of process equipment are actually operating. The most common application is Gamma Scanning. This test is primarily applied, but not necessarily limited to, distillation or separation columns. Gamma scans provide a density profle of the internal process of operating distillation columns and other process vessels. The density profle can be used to diagnose the hydraulic operating conditions of mass transfer devices such as damage to internals, fooding, degree of entrainment or weeping, liquid levels on trays and distributors, liquid distribution through packed beds, etc. The presenter will show case studies where scanning revealed vital process information that helped solve an operating problem, or helped make a revamp/re-design successful. Additional derivatives from gamma scanning of detecting flow distribution patterns will be demonstrated. The ThruVision Scan technique is a specialized horizontal gamma scan used to generate a topographic profile of the internal cross-sectional density of process equipment. This profile is useful for the detailed study of liquid flow distribution through packed columns as well as having applications beyond distillation columns. A hybrid application involving tracers and scanning on fixed bed catalytic reactors will also be discussed.

## 43b The Three-Phase Crystallization of *L*-Menthol from the Enantiomer Mixture

*Lie-Ding Shiau*

Three-phase crystallization (TPC) is a new separation technique which combines melt crystallization and vaporization. Basically, TPC is operated at a triple-point condition, in which the liquid is simultaneously vaporized and crystallized due to the three-phase equilibrium. In essence, the process is continued until the liquid phase is completely eliminated and only the pure solid crystals remain in the feed. Thus, crystal washing is not required since only pure solid crystals remain in the feed and no impurities are adhered on the crystal surfaces at the end of SC. In the previous research, TPC has been successfully applied to separate p-xylene from the mixed xylenes by Shiau et al. (2008, 2013, 2018). TPC is introduced in this work for chiral purification of *L*-menthol from the enantiomer mixture. As TPC is operated in a series of the three-phase transformation conditions, TPC combines melt crystallization and vaporization to produce the *L*-menthol crystalline product and mixture vapor from the mixture melt. The operating pressure for the three-phase transformation during the TPC cooling process is determined based on the thermodynamic calculations. Sweating is applied at the end of TPC to expel the mixture melt remained with the final *L*-menthol product. A model based on the mass and energy balances is proposed to determine the product yield and enantiomeric purity of *L*-menthol for TPC.

References:

1. 1. Shiau, L. D. “Purification of styrene from a styrene/ethylbenzene mixture by stripping crystallization”, *Eng. Chem. Res*, **57**, p6759-6765, 2018.

2. Shiau, L. D.; Liu, K. F. “Investigations into the effects of the cooling rate on stripping crystallization”, Ind. Eng. Chem. Res, 52, p1716-1722, 2013.

3. Shiau, L. D., Wen, C. C. and Lin, B. S. “Separation of p-xylene from the multicomponent xylene system by stripping crystallization”, AIChE Journal, 54(1), p337-342, 2008.

# AUTHOR INDEX

# NOTES

# NOTES

## NOTES